Engineering Marvels

Gingerbread House

Composing Numbers 11–19

Logan Avery

We want to make a gingerbread house.

We can count
what we have.

10 green gumdrops

3 red gumdrops

10 orange candy buttons

4 brown candy buttons

10 pink jellybeans

5 blue jellybeans

10 red and white candy canes

6 green and white candy canes

10 yellow sour candies

9 orange sour candies

We finished our gingerbread house. We had lots of fun!

Problem Solving

Count the chocolate chips for the gingerbread house. Use pictures, objects, and numbers to solve the problems.

1. Use ten frames and numbers to show how many chocolate chips there are.

☐ ones and ☐ more ones

2. Write an equation to match your ten frames.

_____ + _____ = _____

Answer Key

1. 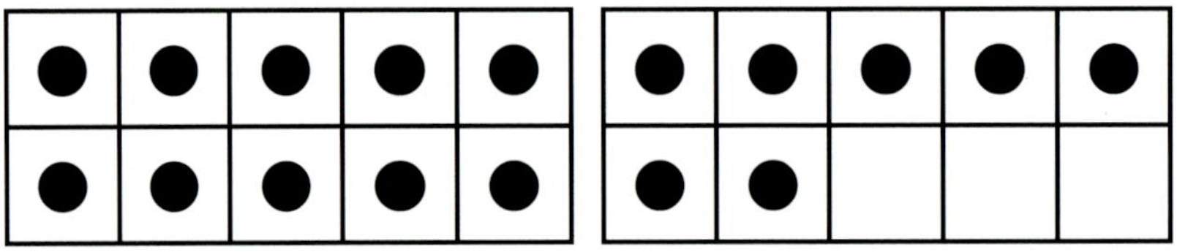

   10 ones and 7 more ones

2. 10 + 7 = 17

Consultants

Nicole Belasco, M.Ed.
Kindergarten Teacher, Colonial School District

Colleen Pollitt, M.A.Ed.
Math Support Teacher, Howard County Public Schools

Publishing Credits

Rachelle Cracchiolo, M.S.Ed., *Publisher*
Conni Medina, M.A.Ed., *Managing Editor*
Dona Herweck Rice, *Series Developer*
Emily R. Smith, M.A.Ed., *Series Developer*
Diana Kenney, M.A.Ed., NBCT, *Content Director*
June Kikuchi, *Content Director*
Véronique Bos, *Creative Director*
Robin Erickson, *Art Director*
Stacy Monsman, M.A., and Karen Malaska, M.Ed., *Editors*
Michelle Jovin, M.A., *Associate Editor*
Fabiola Sepulveda, *Graphic Designer*

Image Credits: All images from iStock and/or Shutterstock.

Library of Congress Cataloging-in-Publication Data

Names: Avery, Logan, author.
Title: Gingerbread house / Logan Avery.
Description: Huntington Beach, CA : Teacher Created Materials, [2019] | Series: Engineering marvels | Audience: Grades K to 3.
Identifiers: LCCN 2017059902 (print) | LCCN 2017060688 (ebook) | ISBN 9781480759633 (e-book) | ISBN 9781425856250 (pbk.)
Subjects: LCSH: Gingerbread houses--Juvenile literature.
Classification: LCC TX771.2 (ebook) | LCC TX771.2 .A94 2019 (print) | DDC 641.86/54--dc23
LC record available at https://lccn.loc.gov/2017059902

Teacher Created Materials
5301 Oceanus Drive
Huntington Beach, CA 92649-1030
www.tcmpub.com

ISBN 978-1-4258-5625-0

Printed by: 51497
Printed in: China